MEGATALL SKYSCRAPERS

and Other City Tech

COOL TECH

World Book, Inc.
180 North LaSalle Street
Suite 900
Chicago, Illinois 60601
USA

For information about other "Cool Tech" titles, as well as other World Book print and digital publications, please go to www.worldbook.com.

For information about other World Book publications, call 1-800-WORLDBK (967-5325).

For information about sales to schools and libraries, call 1-800-975-3250 (United States) or 1-800-837-5365 (Canada).

Library of Congress Cataloging-in-Publication Data for this volume has been applied for.

Cool Tech
ISBN: 978-0-7166-2429-5 (set, hc.)

Megatall Skyscrapers and Other City Tech
ISBN: 978-0-7166-2435-6 (hc.)

Also available as:
ISBN: 978-0-7166-2452-3 (e-book)

Printed in India by Thomson Press (India) Limited, Uttar Pradesh, India
2nd printing November 2021

Credit: © Bill Perry, Shutterstock

STAFF

Editorial

Writer
William D. Adams

Manager, New Content
Jeff De La Rosa

Manager, New Product Development
Nick Kilzer

Proofreader
Nathalie Strassheim

Manager, Contracts and Compliance (Rights and Permissions)
Loranne K. Shields

Manager, Indexing Services
David Pofelski

Digital

Director, Digital Product Development
Erika Meller

Digital Product Manager
Jonathan Wills

Graphics and Design

Senior Designer
Don DiSante

Media Editor
Rosalia Bledsoe

Manufacturing/Production

Manufacturing Manager
Anne Fritzinger

Production Specialist
Curley Hunter

CONTENTS

INTRODUCTION

One of the first things that comes to mind when thinking of a modern city is the soaring skyscraper. As cities expand and more people want to live and work near each other, it makes sense for builders to stack floor on top of floor, creating soaring, delicate towers with commanding views. The tallest of these skyscrapers require extreme engineering solutions to hold themselves up and still maximize floorspace.

But cities are more than just skyscrapers. Vast transit networks, homes, businesses, and factories all combine to make a city what it is. And what a city is today is not what it will be tomorrow. Cities change constantly, and most are growing. More people live in the cities than ever before–more than half of the world's 7.5 billion people. By 2050, that fraction will reach two-thirds.

As more people move to cities seeking opportunities, these cities are at risk of becoming more sprawling, wasteful, and dirty. Haphazard planning slows traffic to a halt. Hastily constructed buildings waste resources and guzzle energy. Cities are responsible for 70% of the world's yearly output of carbon dioxide, a greenhouse gas that is spurring global warming.

Architects, engineers, and urban planners are building a better future for city dwellers. Because so many people live and work in cities, even a minor technological improvement can have a huge impact. With careful planning, cities can be clean, safe, and healthy places to live. Architects can create taller buildings, leaving more space for parks and transportation. Buildings can use less energy and be more responsive to the needs of the people who live and work in them.

How high can skyscrapers climb? What will cities of the future look like? What will this city tech be able to do for you? Read on to find out!

1 MEGATALL SKYSCRAPERS

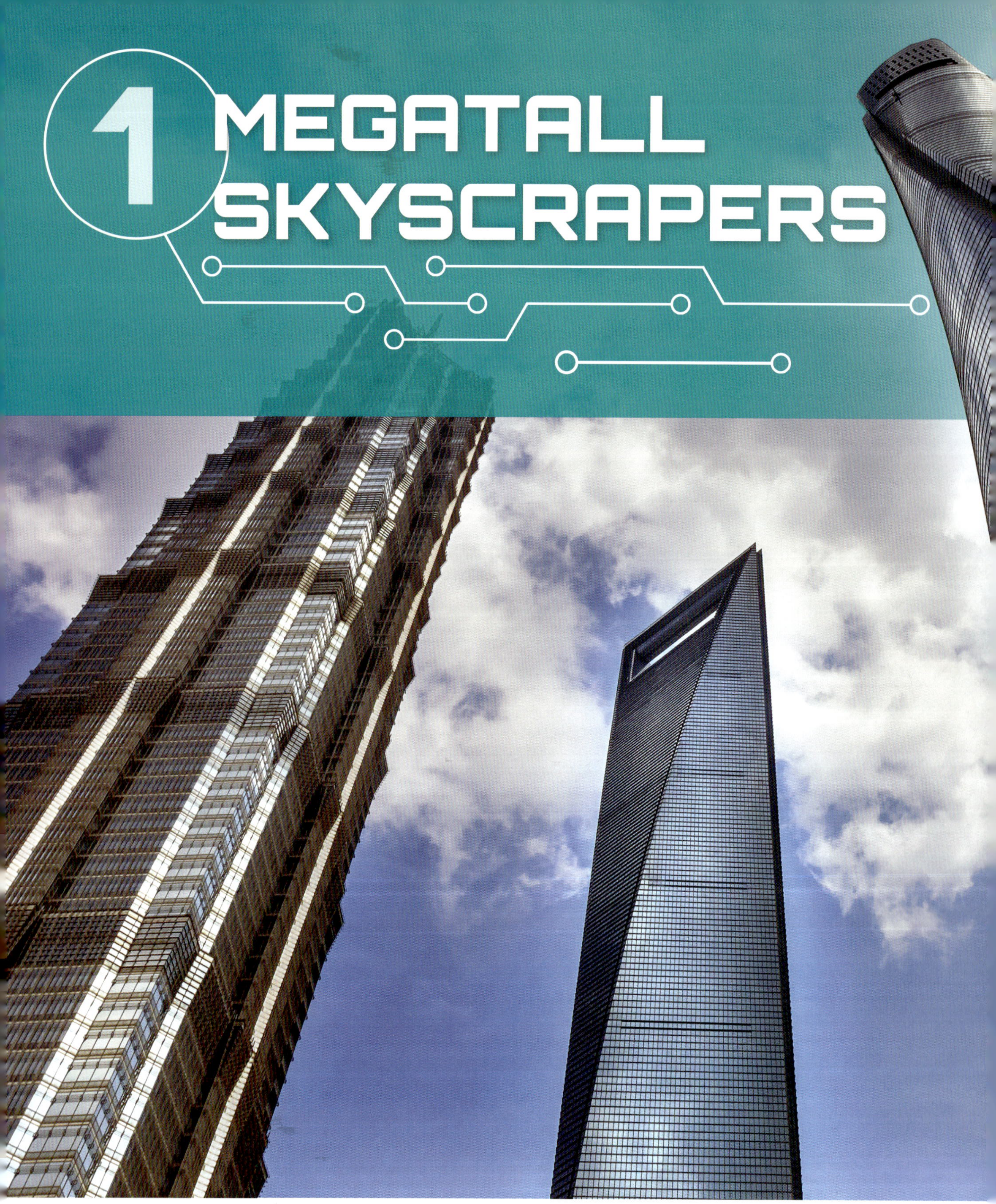

LIFE AT THE TOP

Imagine walking into cavernous lobby, lined with panels of wood, stone, and glass. You enter a plush elevator and press a button. You can feel it accelerating. You can feel the pressure changing in your ears. As the doors open, it seems as if the whole world is laid out in front of you. Large buildings look like models. Cars on the ground below are practically invisible, only identifiable as multicolor streams on the roads below. You look down on helicopters flying above the city. Other cities, regions, and even countries are visible in the distance near the curved horizon. This is the megatall skyscraper.

The Council on Tall Buildings and Urban Habitat defines a megatall skyscraper as any building taller than 1968.5 feet (600 meters). These astonishingly tall structures aren't just like any old building, or even other skyscrapers. They require special thought to be put into every aspect of their design. The materials must be strong enough to hold the weight of the building and its occupants and stand up to wind and even earthquakes. Fresh air, clean water, and other utilities must be provided to each floor. People must be able to access every floor quickly and easily. Also, the structure will tower over everything in the city center, so it should be pleasing to the eye.

THE CLIMB TO THE SKIES

In the 1900's, as cities grew, property values increased, and new building technology was developed, it began to make sense to build up. The skyscraper, the symbol of modern cities, was born of these developments.

Skinny steel. Before the 1900's, most tall buildings were made of brick. Brick was strong but heavy, making it possible for a building to rise only a few dozen stories. The walls and columns of such buildings were extremely thick at the bottom. When steel was developed, **architects** realized they could replace thick stone columns with thin steel ones, increasing rentable floor space. The Home Insurance Building (1884-1885) in Chicago, designed by William Le Baron Jenney, was a well-known "ancestor" of the skyscraper. This 10-story building, which was demolished in 1931, had a metal frame that supported much of its load.

Going up! Around the same time, the American inventor Elisha Graves Otis created the first elevator that had an automatic safety device. If the cable that hoisted the elevator broke, the device prevented the cabin from falling. His invention allowed buildings to explode past the dozen or so floors to which they were previously limited. Elevators also flipped floor rental costs. When the upper floors were just the push of a button away, most tenants preferred the commanding views those floors offered.

The glass curtain. As architects placed their trust in steel, skyscraper walls were relied upon for less and less. Eventually, **engineers** built skyscrapers with steel skeletons that fully supported their weight. The walls were only used for keeping out the elements. Free of structural concerns, architects increased the size of the windows to make bold statements and to increase the amount of natural light let into each floor. This process culminated in the floor-to-ceiling windows of Lever House, completed in 1952. These curtains of windows became the new standard for skyscrapers.

Penthouses are for the poor. During the early history of skyscrapers, height of a building was limited by the number of stairs its occupants were willing or able to climb. In these buildings, the lower floors were more expensive to rent, since it took fewer flights of stairs to climb. Think about it—who would want to live or work on the 12th floor when you had to climb there every day? The higher floors had the cheapest rent, since the renters had to climb many flights of stairs to get there.

THE ILLINOIS

By the 1950's, Frank Lloyd Wright was a living legend. His ground-hugging *prairie style* had transformed American architecture in the first half of the century. But many younger architects had become famous for designing tall skyscrapers. Not to be left out, even as he was pushing 90, Wright created one of his most awe-inspiring designs.

In 1956, Wright hosted an extravagant press conference in which he announced his latest project, the Illinois, named after the U.S. state where it was to be built. It was to be a skyscraper a mile (1.6 kilometers) in height, more than four times taller than the Empire State Building, which was the world's tallest at the time.

Wright planned on a structure unlike that of any other skyscraper. All floors would have been supported by a huge central rod running the full height of the building. The rod would have been sunk hundreds of feet or meters into the ground to stabilize the building. To service the tower, Wright proposed 76 nuclear-powered elevators—each cab five stories all!

Wright never had a customer for the Illinois. He was hoping that his own star power would propel the project forward. When he died a few years later, the plan died as well. It isn't clear that it would have been possible to build the Illinois with the technology available in the 1950's and 1960's. But Wright's amazing design inspired many architects as they built ever taller.

Frank Lloyd Wright, a master of prairie style architecture *(right),* was made famous by his buildings in and around Chicago.

The elegant, tapering form of Wright's proposed skyscraper *(left)* inspired the design of todays megatalls, such as the Burj Khalifa *(background).*

MODERN MEGATALLS

Although Wright's mile-high concept has yet to be equaled, skyscrapers are creeping closer. Only three megatalls have been built so far, but several more are under construction.

Burj Khalifa. The tallest building in the world as of 2019, the Burj Khalifa extends a breathtaking 2,716 feet (828 meters) above the ground. It is located in the United Arab Emirates city of Dubai. Its architects pioneered a type of design called a buttressed core. A thick, concrete core is further supported by three concrete wings, giving the building a Y cross-section. This means that every apartment has stunning views.

Shanghai Tower. This tower is located in the Chinese city of Shanghai. It features an innovative design in which the cylindrical building is nestled inside of a spiraling, tapering, triangular glass tower. This exterior cladding keeps the tower from swaying in the wind and supports large public sky lobbies.

Makkah Royal Clock Tower. This colossal building houses pilgrims to the Muslim holy city of Mecca, in Saudi Arabia. The tower features four 141-foot (43-meter) clock faces. A giant display board shows messages in multiple languages. Special lights on the tower remind Muslims up to 19 miles (30 kilometers) away of prayer times.

ENGINEERING CHALLENGE: MAKING MEGATALLS WORK

Building taller structures by themselves shouldn't be too difficult. Using the buttressed core design pioneered by the Burj Khalifa, engineers think they could build structures more than a mile in height, surpassing even Wright's visionary Illinois. But the taller the building, the more challenges builders and developers have to face to get them built.

Elevators are so essential to skyscrapers that they are effectively designed around them. Many skyscrapers have elevators that stop only at a certain group of floors. For instance, a few elevators might serve floors 11 through 20 in a 50-story building. But these elevators must also stop at the lobby. Their elevator shafts have to run through floors 2 through 9, creating unusable space within those floors.

For tall skyscrapers of about 80 or more stories, this approach doesn't work as well. The lower floors practically fill up with elevator shafts, leaving little room to rent out to businesses. So architects and engineers came up with the idea of sky lobbies. People who want to get to the upper floors of a building first take an express elevator up to a sky lobby. From there, they select from a bank of elevators to reach their destination. This

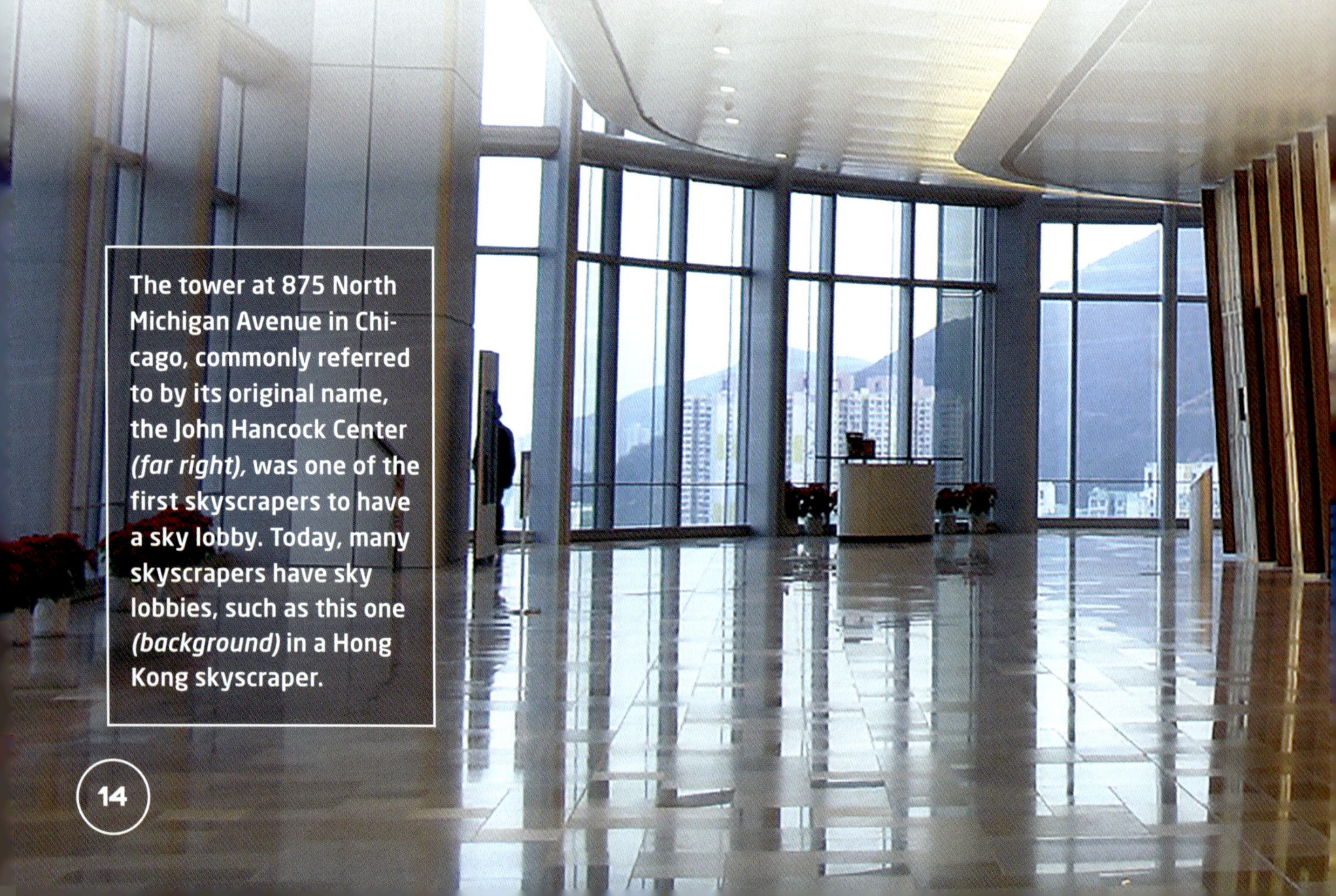

The tower at 875 North Michigan Avenue in Chicago, commonly referred to by its original name, the John Hancock Center *(far right)*, was one of the first skyscrapers to have a sky lobby. Today, many skyscrapers have sky lobbies, such as this one *(background)* in a Hong Kong skyscraper.

approach saves space because the elevator shafts that service the upper floors of a skyscraper no longer have to extend all the way down to the first floor.

But in megatall skyscrapers, even sky lobbies can't prevent elevator shafts from taking up huge amounts of floor space. Plus, the steel cables used to hoist elevators become unmanageable after about 120 floors or so. For skyscrapers to get much taller, architects will have to use new elevator technology.

Such technological hurdles can be overcome. The biggest challenge with building taller skyscrapers is money. The taller the building, the more it costs to build and maintain, which makes it harder for a developer to make money. The Burj Khalifa, for example, actually costs more money to operate than it makes back in rent. But its fame has made the surrounding area, which the same development company owns, very valuable. The company has made enough money developing this area to justify the expense of running the Burj Khalifa. This model is not workable in large, established cities. If architects and engineers can't come up with ways to make skyscrapers taller while increasing floor space or reducing costs, developers won't build them.

THE FUTURE OF MEGATALLS

The Burj Khalifa's days as the world's tallest building are numbered. Another megatall currently under construction will dwarf it. Others will likely come with new ways to push higher.

Jeddah Tower will soar some 3,300 feet (1,000 meters) above the Saudi Arabian city of Jeddah. The building will combine retail space, offices, a hotel, and different kinds of apartments. Guests will take express elevators up to a sky lobby entry floor for each section, where they will then board local elevators to reach their destination. This method keeps the different zones separate, preventing a hotel guest from sneaking into the penthouse, for instance.

Elevators of the future. One company has a solution that might allow megatall skyscrapers of the future to do without dozens of elevator shafts and multiple sky lobbies. In the MULTI elevator design, by the German company ThyssenKrupp, elevator cabins travel up a magnetic track. No longer bound by cables, MULTI cabins can change shafts using a turntable device—even as the cabin stays perfectly upright. With this setup, several cabins may run in a single shaft. If one cabin is stopped, others in that shaft can move to different shafts to complete their journeys. Cabins could even travel to other buildings via special bridges or directly to public transit stations.

Merdeka PNB118. This megatall will be located in Kuala Lumpur, the capital city of the Southeast Asian country of Malaysia. Kuala Lumpur already boasts the twin supertall Petronas Towers, among the tallest buildings in the world. What's in a name? *Merdeka* means *independence* or *freedom* in Bahasa Indonesia (Indonesia's official language), in reference to the country's establishment in 1945. PNB stands for the developer, Permodalan Nasional Berhad. And 118 is the number of floors that will make up the building.

2 SMART CITIES

LIVE SMARTER, NOT HARDER

Imagine being in a car sitting at an intersection, waiting for a red light to change. No other vehicles are moving across the intersection, so you're just waiting for no reason. This minor annoyance incurs a real cost in the form of lost time and lost fuel. But it's also a major concern for the city in which this waste occurs. Such minor inefficiencies compound into major traffic headaches. And all the extra traffic creates air pollution and noise pollution.

What if traffic lights could figure out when people were needlessly waiting and change their signals to accommodate them? And what if they could communicate with each other so you got a series of green lights as you drove to your destination? Such systems are being installed in cities across the world.

This is just one way that cities are getting smarter to help their residents live better lives. New technology is enabling city officials to collect more data about the city and respond to them. Cities with these systems in place are called smart cities.

BLAST FROM THE PAST

Historically, cities have mostly been products of convenience. People built them in places that were easy to defend from invading armies or near connections between modes of transportation. Buildings sprung up around old trails, and more streets radiated out from them as a settlement grew. Many cities were a tangle of roads, rails, buildings, and drainage ditches.

Urban planning. As long as cities have existed, people have imagined designing them to make them better. At first, people trying to improve cities usually relied on the commands of rulers or their own ideas about cleanliness and order to inform their designs. But starting in the 1900's, people began using more scientific approaches, using information on how people live in cities to figure out what improvements should be made to them. This field is called urban planning.

The man with the plan. In 1909, funded by Chicago businesses and wealthy citizens, the architect and urban planner Daniel Burnham released a comprehensive plan for Chicago. It called for broad, tree-lined boulevards; large parks; and awe-inspiring public buildings. While many of the specific recommendations of the Burnham plan were never followed, many of its general ideas and predictions about the expansion of the city over the next 100 years proved to be correct. It remains an important document in the history of **urban planning.**

Urban utopias. Sometimes, urban planners or other visionaries sought to tame the chaos of cities by designing new urban centers. Very few—if any—of these planned cities lived up to their creators' visions. But many still exist today and provide their citizens with a pleasant living and working experience.

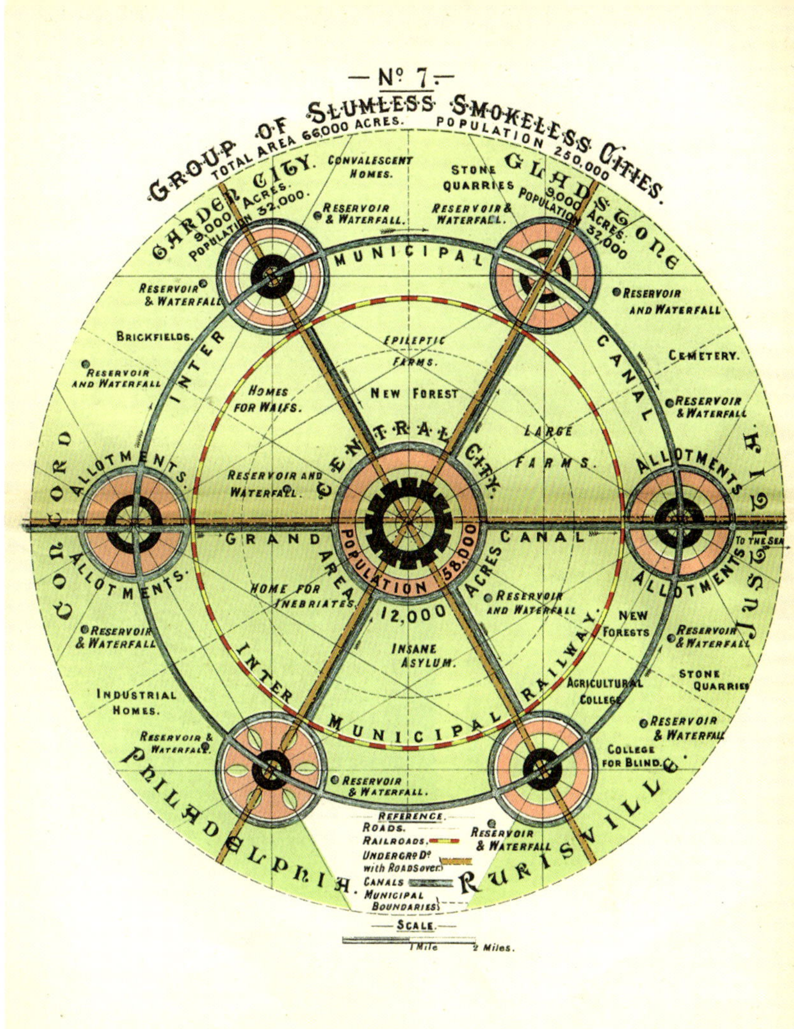

British urban planner Ebenezer Howard published his vision for a "garden city" *(right)* in 1902. Swiss-born architect Charles-Édouard Jeanneret envisioned buildings as villages in his modern designs *(below)*.

EPCOT

The American animator Walt Disney built a media empire with charming characters, vivid animations, and exciting stories. In 1955, he invented the modern theme park with Disneyland in California. But he wasn't satisfied. In Florida, he purchased land for a futuristic planned city that he called the Experimental Prototype Community of Tomorrow, or EPCOT.

Walt Disney envisioned EPCOT as the main piece of a huge Florida development that would also contain a theme park, an industrial park, and an airport. About 20,000 people were to live in EPCOT. He imagined EPCOT and the industrial park as ever-changing laboratories for the development of urban technology.

EPCOT was to have a radial layout, like a wheel with spokes. The downtown, containing a huge hotel and convention center, would have been in the middle of the wheel. Around it would have been offices and businesses. Some of the streets would have had themed restaurants, shops, and architectural elements inspired by different countries of the world. Farther along the spokes would be sprawling, suburban developments.

The spokes would not have been roads, but tracks. People would have walked from their homes to board small, automated trains called PeopleMovers, which would have taken them to the center of the city. From there, they would have walked to their jobs or taken a larger and faster monorail to the industrial park or theme park, where more PeopleMovers would have brought them to their jobs.

The original design for EPCOT was for a futurisitc city based on a radial layout *(right)* that would house about 20,000 people.

Walt Disney *(left)* imagined EPCOT as an ever-changing laboratory for urban planning and the design of future cities.

In EPCOT, cars and trucks would have been banished to run below the city on two separate, underground levels.

After Walt Disney died in 1966, none of the other top executives in the company wanted to build his ambitious city, so the project was cancelled. Only the theme park, now called the Magic Kingdom, was built. Later, the company reimagined EPCOT as another theme park, which opened as Epcot Center in 1982. Some of the theme park's attractions, such as the World Showcase, are directly inspired by Disney's ideas for EPCOT.

SMART CITY EXPERIMENTS ACROSS THE GLOBE

Recently, some urban planners have focused on gathering more data on how cities function and using those data to make informed policy decisions. The goal: cities that work better. This usually requires hundreds of pieces of such equipment as cameras and **sensors.** A city that has this data-gathering infrastructure is said to be a smart city.

Array of Things. The Array of Things is a collection of more than 100 sensor nodes spread throughout Chicago. These nodes detect such things as air quality, temperature, and even vibrations. The data are freely available to city officials and researchers. Proponents of the initiative hope to install 500 nodes in the city and expand to other cities worldwide. Part of the success of Chicago's Array of Things is that it is open source and is designed to be transparent and anonymous. A node tracks and counts the cars and pedestrians that pass under it, but it sends only anonymous data, such as the number of pedestrians and directions in which they are moving, to a central hub. Once it has done this, it erases the original footage. This way, citizens can rest assured Array of Things's nodes can't be co-opted to spy on them.

Smart cities require data on how people move, shop, work, and play. Urban planners can use this data to make useful designs.

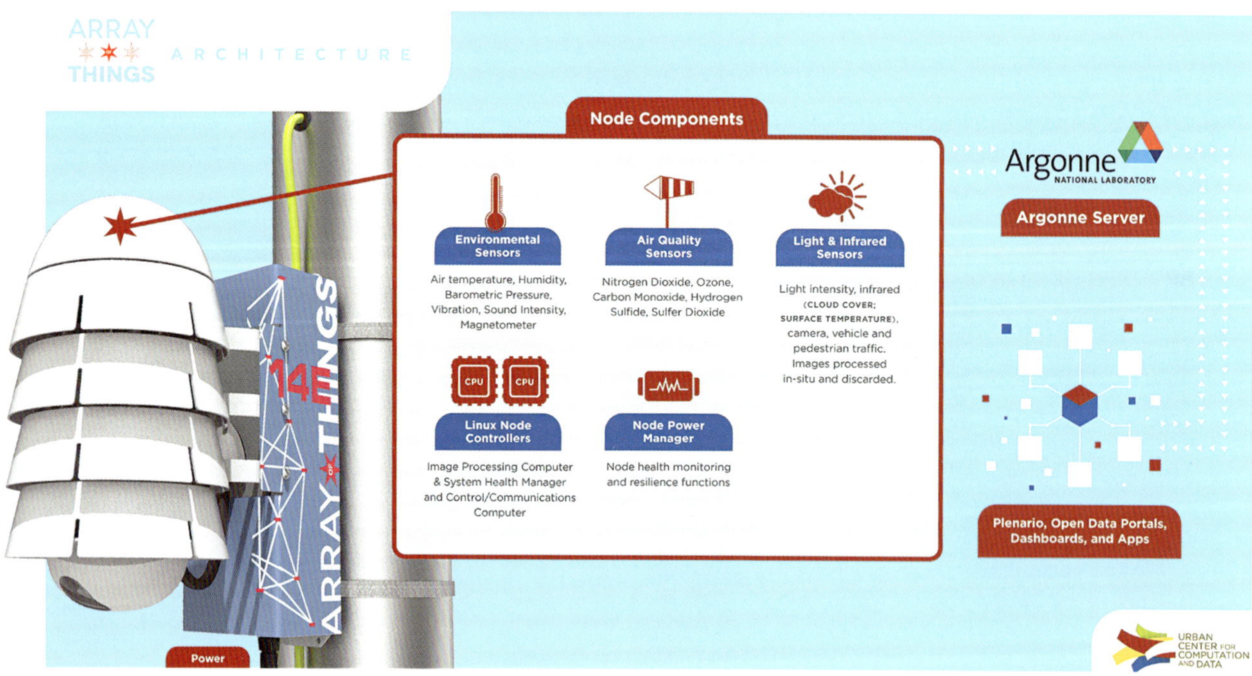

Singapore. The Southeast Asian island city-state of Singapore is one of the smartest cities in the world. The government has built a detailed three-dimensional model of the entire city. City planners can use the model to determine how new buildings, roads, and other projects will affect traffic flow. Because the whole country is so small, the government plans to extend the initiatives across the country.

Sidewalk Toronto. Google sibling company Sidewalk Labs is partnering with the city of Toronto to redevelop a portion of the city's waterfront area, called Quayside. Sidewalk Labs plans to build a smart neighborhood from the ground up, with flexible streets, modular green buildings, and ubiquitous sensors.

THE FUTURE OF SMART CITIES

Many improvements gained from smart city infrastructure won't be obvious. Smart city technologies won't impact a city's skyline, for example. But they will help make cities better places to live.

Less traffic. The biggest benefit from smart cities will be reduced traffic. Connected stoplights will be able to predict traffic patterns and change lights to ease congestion. Many cities already use similar technology to allow buses and emergency vehicles to change stoplights. Such signal timing systems can pay for themselves many times over in the form of increased productivity.

Cleaner, greener city. Smart city technology could greatly reduce the amount of resources a city uses. Smart water and power meters help customers understand their usage patterns and suggest ways to save. Sensors along power and water grids can detect leaks and breakdowns. Utility crews can then quickly fix the problem before more water or power is wasted.

Better urban design. Urban planners are trying to make cities better places for everyone, but they have few data with which to work. Many of their ideas are still hunches, and those that are put into practice don't always pan out as expected. With ubiquitous sensors and powerful computers, urban planners can better study cities, figure out what works and what doesn't, and create plans to promote successful development.

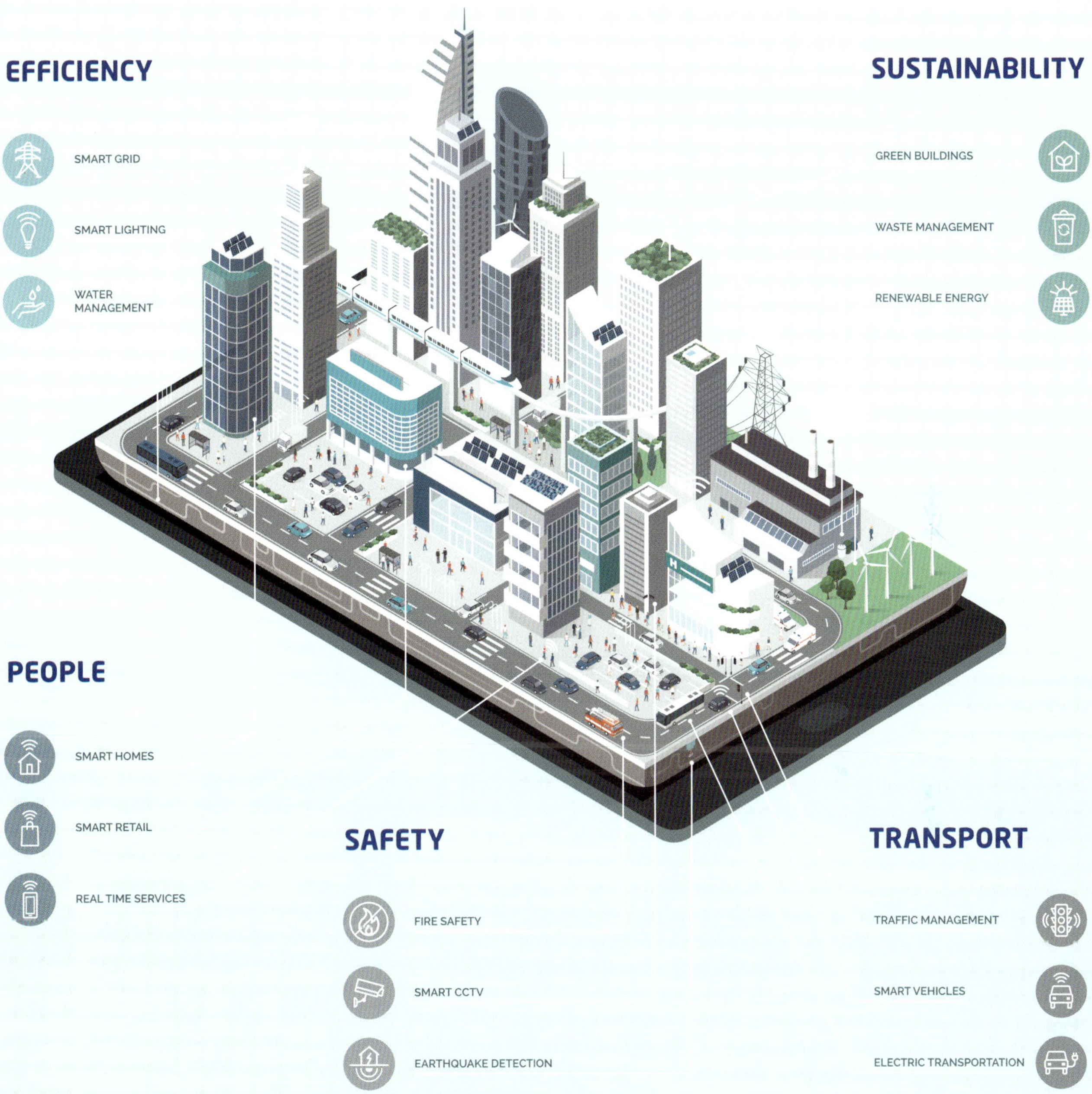
EFFICIENCY
SMART GRID
SMART LIGHTING
WATER MANAGEMENT
SUSTAINABILITY
GREEN BUILDINGS
WASTE MANAGEMENT
RENEWABLE ENERGY
PEOPLE
SMART HOMES
SMART RETAIL
REAL TIME SERVICES
SAFETY
FIRE SAFETY
SMART CCTV
EARTHQUAKE DETECTION
TRANSPORT
TRAFFIC MANAGEMENT
SMART VEHICLES
ELECTRIC TRANSPORTATION

3 GREEN BUILDINGS

IT'S NOT EASY BEING GREEN

Think about things that produce a lot of pollution. Cars, factories, and coal-fired power plants probably come to mind. But what about residential and commercial buildings? You might not use very much energy in your home compared to a factory, but there are millions and millions of homes like yours. And the energy used in all those buildings adds up. In fact, homes and business account for about 39% of all energy consumed in the United States.

With the looming threat of global warming, energy efficiency is a huge priority. We can make big reductions in the amount of greenhouse gases released into the **atmosphere** by taking careful stock of how we design, build, maintain, and repair buildings. This is the science of green building.

Balancing all these variables with other building demands isn't easy. In fact, engineers and architects are still coming up with ways to reduce the environmental impact of structures and measure their energy performance. But with modern building techniques, buildings can be stylish, comfortable, and cheap to maintain. Not only are green buildings good for the environment, but they can also save money and make people happier, healthier, and more productive. For example, well-ventilated, well-lit buildings (which green buildings often are) make people who live and work in them healthier and more productive.

GREEN BUILDING MATERIALS

Most buildings are designed to last for decades, or even centuries. So why should **architects** worry about the environmental profile of the materials it is made of? As buildings become more efficient in how they are heated, cooled, and lit, the greenhouse gases and pollution put out during their production take up a greater percentage of lifetime emissions—as much as 40%.

Embodied energy. All building materials, whether wood, glass, brick, or drywall, take energy to create and move to a building site. This is called embodied energy. As buildings use less energy to heat and cool, the amount of embodied energy in a building's materials has become a greater part of the energy it will use in its lifetime. Therefore, limiting embodied energy is important.

Other considerations. There are more things to think about than just embodied energy. Glass is not a good insulator, but very few people would want to live in a windowless house. Other materials might have a low embodied energy, but might contain **volatile organic compounds (VOC's).** VOC's give off toxic gases, which can cause respiratory illnesses and may be linked to some forms of cancer.

A traditional thatch-roofed cottage made from local materials *(left)* has a naturally low embodied energy cost. Reclaiming and reusing building materials *(upper right)* helps lower the embodied energy cost of new construction. Repurposing an old factory into apartments *(lower right)* is almost always a more efficient use of energy and materials.

Reclaimed building materials. Using materials salvaged from old buildings is a green way to add character to a structure. New versions of some hand-crafted home elements, such as an intricately-carved fireplace mantel, are practically impossible to find today, since both the skills and materials needed to make them have become hard to come by.

The greenest building material is the one that's already in place. Should a corporation tear down its old blocky office building and replace it with a sleek, green building to help the environment? Probably not. Studies show that it's almost always better for the environment to reuse an existing building than to tear it down and build a new one. With careful planning, a redesigned building can almost match the environmental performance of a replacement building. Rehabilitating uses far fewer materials that would be required in a complete teardown.

WOOD: GREEN BUILDING MATERIAL OF THE FUTURE?

Concrete and steel became the materials of cities and skyscrapers, but they have sky-high embodied energies. Harvesting, transporting, and processing timber takes much less energy. In fact, timber can *sequester* carbon (store it for a long time). Trees take up carbon dioxide from the atmosphere and use sunlight to convert it into energy and tissues, such as leaves and wood. When the tree is cut down and turned into timber for a building, the carbon in the wood is prevented from returning to the atmosphere, as it would have if the tree died in the forest. A new tree grows in the place of the tree that was cut down, absorbing more carbon dioxide.

Why did so many architects switch to concrete and steel? Wood fell out of favor in dense urban environments not because it wasn't strong enough, but because of its flammability. Wood-framed buildings burn easily because there is a large contact area between the surface of the long, narrow pieces of framing wood and the source of the fire. After fires ravaged cities for decades, many adopted building codes that promoted concrete and steel over wood for larger structures.

Large pieces of wood—called mass timber—react differently to fire than framing wood. The outer layers char, temporarily protecting the inside. This allows a mass timber building to stay together for a long time when on fire. Mass timber is just as strong as steel, but lighter. Mass timber made of multiple layers of wood running in different directions (called **cross-laminated timber**, or CLT) provide excellent strength in all directions.

Architects are looking to build large buildings, even skyscrapers, out of mass timber. Many cities are still suspicious of timber construction, having been burned in the past by terrible fires fueled by wooden structures. But the success of the first modern mass timber buildings, coupled with increasing concern about climate change, may lead governments to change their stances on wooden buildings.

Architects are looking to build more buildings with wood. Designs for skyscrapers made of wood *(right)* have even been proposed!

GREEN BUILDING DESIGN

A building's greenness is determined by more than just the materials it is built from. The building's design, its fixtures and equipment, and how people use and maintain it all affect energy use.

Active green design. Active green design components use or enable some energy to allow the structure to use less energy overall. Rooftop solar panels, for example, are an active green design component. With solar panels hooked up to the home's electrical grid, the heat of the sun can be used to cool the building if necessary. Energy-efficient heating and cooling devices, such as furnaces and fans, are also important. In the United States, the government-backed program EnergyStar labels products that are energy efficient. Fans can make rooms more comfortable by spreading warm or cool air around a room. This can allow people to raise or lower the thermostat.

New green buildings feature advanced materials and innovative designs to maximize energy efficiency and comfort. Existing homes can be upgraded to get more green!

Passive green design. Passive green design calls for buildings that make the best use of such natural elements as sunlight, wind, plants, and rainwater to limit the use of energy and other resources from outside sources. An easy thing to do is to use awnings or landscaping to control sun exposure. During the spring and summer, leafy trees or awnings block sunlight from warming a building through sun-facing windows. But in fall and winter, the leaves fall from the tree or the path of the sun sinks below the awnings, allowing sunlight to warm the building.

Simple upgrades. All of these design considerations are great for **architects,** but what about homes and other structures that have already been built? Simple upgrades can have a major impact on your energy bill. Heated or chilled air can flow in and out of tiny seams and holes in a house. Finding and sealing these holes can go a long way towards making a home more comfortable. Areas that can't be caulked, such as around doors and windows, can be sealed with weather-stripping. Upgrading insulation within a wall is a bigger job, but can also save money and energy.

4 VERTICAL FARMING

GROWING UP, NOT OUT

More than half the world's population lives in urban areas. What if you could grow the food that they eat in the city? Land is expensive in cities, so traditional farms couldn't possibly exist there. But if people could grow stacks of crops in a warehouse, or even in a skyscraper, city dwellers could have easy access to fresh, delicious produce. This idea of growing crops in multiple layers is called vertical farming.

For the last 75 years or so, most crops have been grown on huge farms. These farms use large amounts of mechanization in planting, fertilizing, and harvesting crops. They produce lots of food, but they are at risk from bad weather and pests. They are also often located far from where their crops are consumed. In part because they travel so far, most fruits and vegetables have been bred not to taste better, but to last longer. A fruit or vegetable might spend weeks or even months in storage and transit before it reaches a restaurant or supermarket.

Because they could be located so close to where their produce will be sold, vertical farms could focus on freshness and quality. The food doesn't have to be trucked into the city, which reduces both greenhouse gas emissions and traffic. Vertical farmers can more carefully protect their crops from pests and disease, effectively eliminating the need for harmful pesticides. There's no polluting fertilizer runoff into the water supply, either. When it comes to a vertical farm's advantages, the sky's the limit!

HOW VERTICAL FARMING WILL WORK

Vertical farming is still in its infancy. It has a long way to go before it hits the mainstream. But most proposals tackle the challenge of growing food indoors in similar ways.

Vertical farms are already popping up in several large cities. These farms use advanced indoor hydroponic and lighting systems to produce vegetables for sale.

Hydroponics. Since vertical farms are indoors, the plants have to be provided water. Most vertical farms use a system called **hydroponics**, where plant roots are placed in a bath of nutrient-rich water. It is easier to filter and add nutrients to water than soil. Hydroponic systems also protect plants from disease, since the microbes that cause plant diseases often live in soil.

Light. Sunlight, a given in traditional farms, is far from guaranteed in vertical farms. Each growing layer blocks the layer above it. Some vertical farms use mirrors or reflectors to get light to lower layers. But in most designs, the crops will need at least some artificial light. Light-emitting diodes (LED's) are what have made vertical farming possible. They use far less energy and produce much less heat than incandescent or fluorescent bulbs. Engineers can tune the LED's to emit exactly the wavelengths of light that the crops need.

Energy use. The need for artificial lighting puts vertical farms at an economic and environmental disadvantage. Today, most electricity is produced by burning fossil fuels, which releases such greenhouse gases as carbon dioxide. Even with energy-efficient LED's, the amount of energy needed to grow crops indoors adds up. A study found that the cost of energy needed to grow enough wheat to make a loaf of bread would be $11!

THE PLANT

One project uses vertical farming as one part of a complete system of sustainable development. The Plant is a cooperative business project located in an old meatpacking building in Chicago. The Plant also focuses on research, education, and community outreach. It is striving to create closed-loop systems, where waste from one tenant or building process is used for something else.

The Plant uses a method called **aquaponics** in its vertical farms. Aquaponics perfectly showcases its closed-loop mission. Fish called tilapia are farmed in huge tanks in the basement of the building. Fish waste is a natural plant fertilizer. But the fish will die if the waste builds up in their tanks. So pumps feed the dirty water into hydroponic vertical farms, where plants absorb the fish wastes dissolved in it. Then the clean water is pumped back into the tilapia tanks. Both the crops and the fish are harvested for use in the Plant's restaurants.

The Plant isn't content with stopping there. The heart of the project will be an **anaerobic** digester: a giant chamber where microbes can turn organic wastes into fertilizer and

natural gas. The fertilizer will be used to nourish the aquatic plants that are fed to the tilapia. The natural gas will fuel an on-site generator that will provide power to the Plant.

The Plant seeks out tenants that will work with its mission. Within it, different shops make coffee, kombucha, and baked goods, often using materials from the Plant's farms or other systems and feeding useable waste back into those systems. The Plant hopes to be a model for sustainable development that the rest of the world can follow.

When the Plant's anaerobic digester *(left)* becomes operational, it will be the heart of the facility. There, organic waste that can't be reused in other ways will be converted into fertilizer and fuel.

5 SMART HOMES

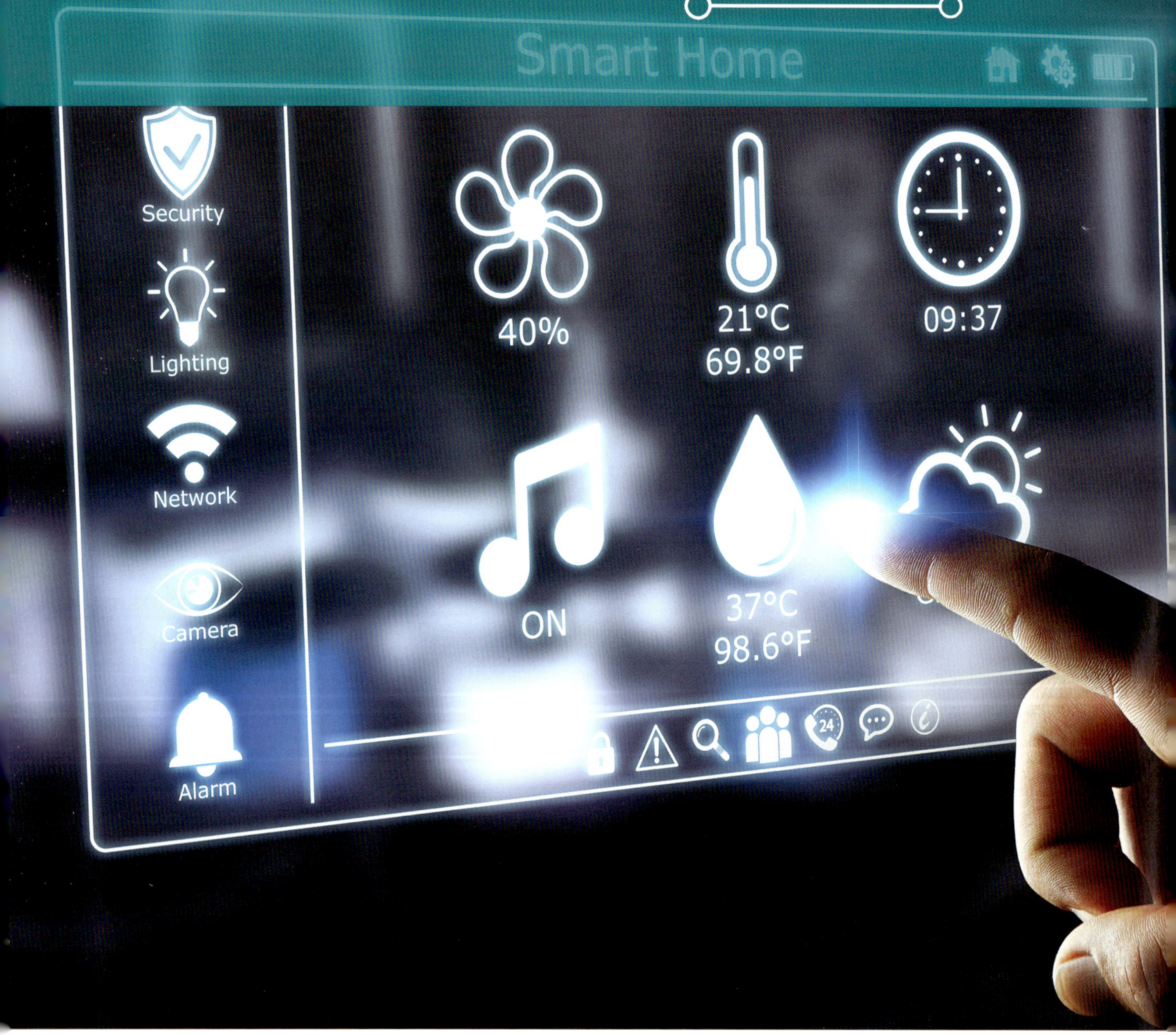

WELCOME TO MY HOUSE

Taking care of a home is a full time job. There's always something that needs fixing or a chore that needs doing. Today, we get some help from technology. Machines wash our clothes, clean our dishes, and sweep our floors. But many of these items operate independently from one another. You can't tell your house to open the windows when it gets to be a certain temperature outside, for example.

That is starting to change. More and more appliances are being connected in homes to help us get more things done. Homebuilders are incorporating more smart technology into different components of the home, such as doors and windows. With smart home technology, you can make your home work for you.

SMART HUBS

For years, smart home technology was a jumble of competing systems, none of which worked together. They used wires hidden in walls to connect to hubs, so it was prohibitively expensive to install or upgrade them in homes that had already been built. Then, in the mid-2010's, companies developed **smart speakers.** These systems, imbued with voice-activated personal assistants, became hugely popular. They quickly became the hub of a smart home. A homeowner's spoken command can turn on the lights, lower the temperature, or arm the security system.

Connected devices. A smart home is only as useful as the smart devices connected to it. Common connected devices include lights, thermostats, and security-camera-equipped doorbells. Once the hub has received a command, it must relay it to the proper device. Hubs deliver commands through such wireless protocols as Wireless internet (WiFi), Bluetooth, Zigby, and Z-Wave.

Hacking and privacy. Smart homes rely on many individual pieces of hardware. Many of these smart devices are always turned on. Most smart speakers, for example, record everything their microphones pick up, not just the commands they are given. If the software of each smart device in a home is not securely designed, its residents could be in danger of spying, blackmail, and theft. But people have to worry about more than just individual hackers. Most smart speakers send recordings to their manufacturers. These data are used to improve smart speaker performance, but users must trust the company with their personal conversations. The company might fall prey to hackers, give the conversations to governments or law enforcement, or decide to study the conversations to sell ads or products.

More energy use? Some experts worry that smart homes will enable more, not less energy use. For example, people may preheat or precool their smart home before they return, or do the same with their smart-home-connected car before they leave. The many gadgets associated with smart homes take energy and materials to produce, use electricity even when they are not in use, and are hard to recycle. In one electrifying study, a connected LED light bulb was found to consume more energy when it was turned off than when it was turned on!

Newly constructed homes may be designed and built with smart systems installed. Older homes may have smart systems added to them by installing smart speakers.

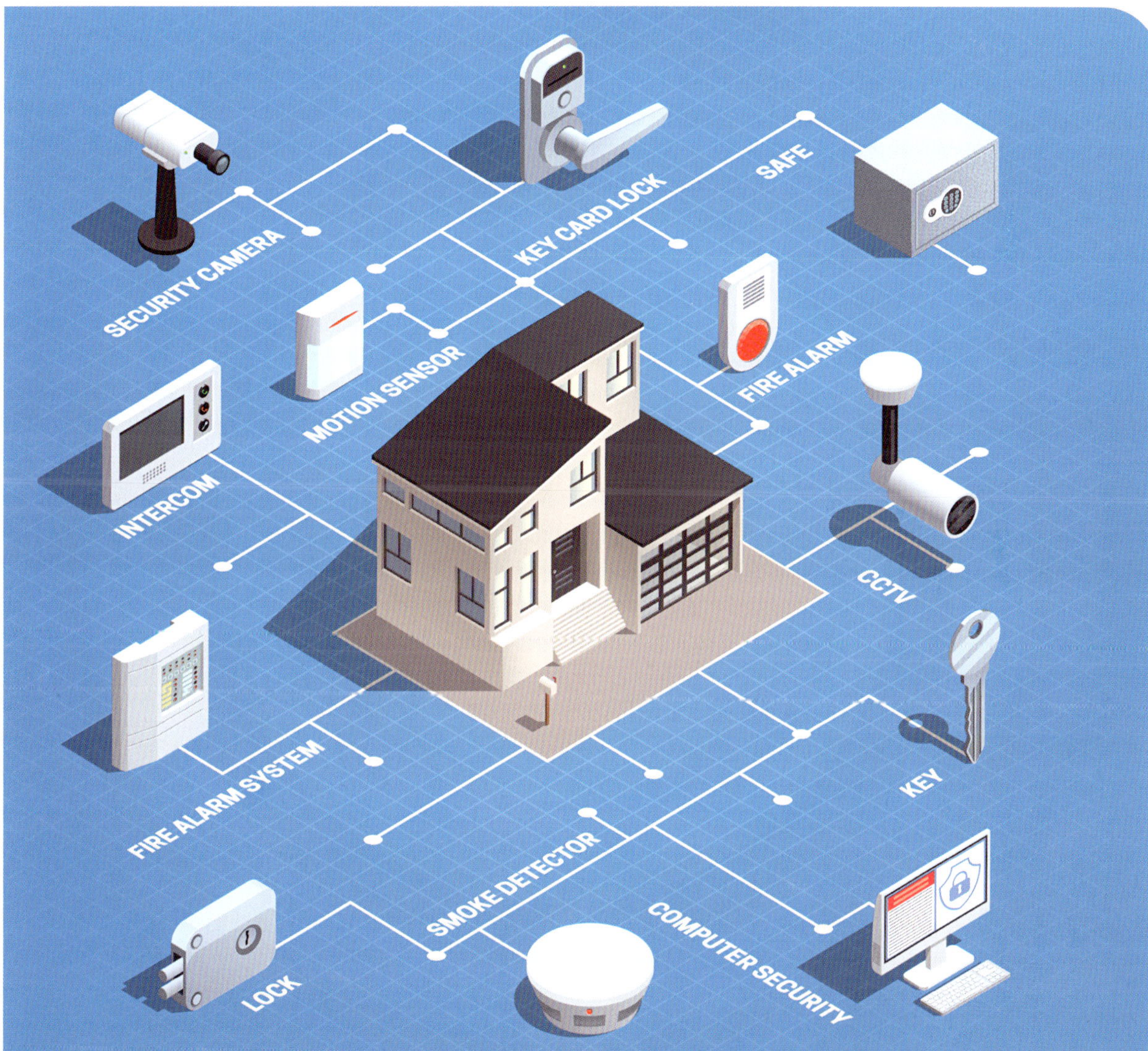

GLOSSARY

anaerobic living, growing, or taking place where there is no available oxygen.

aquaponics a type of hydroponics in which fish waste nourishes plants.

architect a person who designs and lays out plans for buildings.

atmosphere the mass of gases that surrounds a planet or other heavenly body.

cross-laminated timber (clt) a wood building material made of sheets of long planks of wood glued together so that the planks in each sheet run perpendicular (at a 90 degree angle) to each other.

engineer a professional who plans and builds engines, machines, roads, or the like.

hydroponics the science and technology involved in growing plants without soil.

sensor a device that takes in information from the outside world and translates it into code.

smart speaker a small computer that has a microphone and speaker but usually no screen. A smart speaker is loaded with **voice assistant** software that allows its user to find information on the internet and control connected devices with voice commands.

urban planning the process for guiding the development of cities and towns.

voice assistant a voice-activated, interactive program that can be used to control connected devices, find information on the internet, and perform other tasks.

volatile organic compound (VOC) an unstable substance that breaks down over time and gives off small amounts of toxic gases.

INDEX

ACKNOWLEDGMENTS

5	© Adrian Smith + Gordon Gill Architecture
6-7	© Bill Perry, Shutterstock; © Simon Song, South China Morning Post/Getty Images
8	© Elena Miv, Shutterstock; Public Domain; Library of Congress; © Bettmann/Getty Images; © Sean Pavone, Shutterstock
10-11	© Naufal MQ/Shutterstock; © Marek Lipka-Kadaj, Shutterstock; © From Original Negative/ Alamy Images
12-13	© Shutterstock
14-15	Baycrest (licensed under CC BY-SA 2.5); © John Kershner, Shutterstock
16-17	© Adrian Smith + Gordon Gill Architecture; © Fender Katsalidis Architects; © ThyssenKrupp
18-19	© Bavo Studio/Getty Images; © Ross Helen, Shutterstock
20-21	Public Domain; © Chicago History Museum/Getty Images; © Toronto Star Archives/ Getty Images; Public Domain; © Andia/Alamy Images
22-23	© Atlantide Phototravel/Getty Images; Library of Congress; © Consolidated News Pictures/ Getty Images; © Michel Baret, Getty Images
24-25	© Array of Things; © Imran Ahmed, Dreamstime; © Sidewalk Labs
26-29	© Shutterstock
30-31	© Eye Ubiquitous/UIG/Getty Images; © John Ewing, Portland Press Herald/Getty Images; © Adrian Sherratt, Construction Photography/Avalon/Getty Images; © Leeds Fotografica/ Shutterstock
32-33	© wk1003mike/Shutterstock; © PHG Pictures/Shutterstock; © Kilian O'Sullivan, View Pictures/UIG/Getty Images; © Perkins + Will
34-35	© Radovan1/Shutterstock; © Studio Gang; © Raymond Boyd, Getty Images; © Andrey Popov, Shutterstock
36-37	© Xinhua/eyevine/Redux Pictures; © Aisyaqilumaranas/Shutterstock
38-39	© Shutterstock
40-41	The Plant; Bubbly Dynamics
42-45	© Shutterstock